ζ Zeta

Decimals and Percents

Tests

Math·U·See®

1-888-854-MATH (6284) - mathusee.com
sales@mathusee.com

Zeta Tests: Decimals and Percents

©2012 Math-U-See, Inc.
Published and distributed by Demme Learning

mathusee.com

1-888-854-6284 or +1 717-283-1448 | demmelearning.com
Lancaster, Pennsylvania USA

ISBN 978-1-60826-077-5
Revision Code 0120-B

Printed in the United States of America by LSC Communications US, LLC
 2 3 4 5 6 7 8 9 10

For information regarding CPSIA on this printed material call: 1-888-854-6284
and provide reference #0120-12292020

LESSON TEST

Rewrite each number without an exponent.

1. $1^4 =$ ____

2. $10^2 =$ ____

3. $13^2 =$ ____

4. $3^3 =$ ____

5. $12^2 =$ ____

6. $7^2 =$ ____

Write the missing exponent.

7. $5^{\underline{}} = 25$

8. $4^{\underline{}} = 64$

9. $2^{\underline{}} = 16$

10. $6 \times 6 \times 6 \times 6 = 6^{\underline{}}$

11. $15 \times 15 = 15^{\underline{}}$

12. $9 \times 9 \times 9 = 9^{\underline{}}$

Express 8^3 in at least two ways.

13.

14.

Find the fraction of the number.

15. $\frac{1}{4}$ of 96 = ___

16. $\frac{3}{8}$ of 88 = ___

17. $\frac{7}{10}$ of 100 = ___

18. Jessica practiced the piano for 3/4 of an hour. For how many minutes did she practice the piano?

19. One eighth of the apples in the bag were rotten. If there were 24 apples in the bag, how many were rotten?

20. Nine tenths of the people in the room thought the idea was very good. If there were 20 people in the room, how many thought the idea was very good?

Fill in the blanks.

1. 10^4 = _____

2. 10^2 = _____

3. 10^3 = _____

4. $10 = 10$ ——

5. $1,000,000 = 10$ ——

6. $1 = 10$ ——

Write in expanded notation.

7. $125 =$ _____

8. $3,932 =$ _____

Write in exponential notation.

9. $9,028 =$ _____

10. $16,090 =$ _____

Express in decimal notation.

11. $8 \times 10,000 + 3 \times 1,000 + 2 \times 100 =$ _____

12. $2 \times 10^3 + 6 \times 10^2 + 8 \times 10^1 =$ _____

Fill in the blanks.

13. $8^2 =$ _____

14. $2^3 =$ _____

15. $5\text{——} = 125$

Fill in the missing numbers to make equivalent fractions.

16. $\dfrac{1}{8} = \dfrac{}{} = \dfrac{}{24} = \dfrac{4}{}$

17. $\dfrac{4}{5} = \dfrac{8}{} = \dfrac{}{} = \dfrac{}{20}$

18. Max the pug slept for 3/6 of the 24-hour day. For how many hours did Max sleep?

19. For how many hours was Max awake during the day mentioned in #18?

20. Rachel practiced 1/2 of an hour on Monday, 1/3 of an hour on Tuesday, 1/4 of an hour on both Wednesday and Thursday, and 1/2 of an hour on Friday. Find how many minutes she practiced each day and then find how many minutes she practiced in all during the week.

Write in exponential notation.

1. 34.6 = _____

2. 0.124 = _____

Write in decimal notation.

3. $5 \times 10^3 + 4 \times 10^2 + 1 \times 10^1 + 7 \times 10^0 + 4 \times \dfrac{1}{100}$ = _____

4. $1 \times 10^3 + 8 \times \dfrac{1}{10^1} + 7 \times \dfrac{1}{10^2} + 5 \times \dfrac{1}{10^3}$ = _____

Fill in the blanks.

5. \$3.05 = 3 dollars and 0 _____ and 5 _____ = ____ + ____ + ____ = ____

6. \$6.41 = 6 _____ and 4 _____ and 1 _____ = ____ + ____ + ____ = ____

Rewrite each number without an exponent.

7. 10^3 = ____   8. 1^7 = ____

9. 6^2 = ____   10. 11^2 = ____

Fill in the missing numbers to make equivalent fractions.

11. $\dfrac{2}{9} = \dfrac{}{18} = \dfrac{}{} = \dfrac{8}{}$

12. $\dfrac{1}{7} = \dfrac{2}{} = \dfrac{}{} = \dfrac{}{28}$

Simplify each fraction to lowest terms.

13. $\dfrac{3}{12} = $ ——

14. $\dfrac{2}{50} = $ ——

15. $\dfrac{8}{64} = $ ——

16. $\dfrac{15}{90} = $ ——

17. Jenny has three pennies, six dimes, and four dollars. Express this amount in decimal form.

18. Wayne has 5/10 of a dollar. How many cents does he have?

19. Fifty out of one hundred, or 50/100, of the chocolates have nuts. Express the part of the chocolates that have nuts in lowest terms.

20. Using unit blocks, Rose built a square with five blocks on a side. Use an exponent to tell how many blocks she used.

LESSON TEST

Add the decimal numbers.

1. $\begin{array}{r} 6.7 \\ + 5.4 \\ \hline \end{array}$

2. $\begin{array}{r} 2.0 \\ + 0.2 \\ \hline \end{array}$

3. $\begin{array}{r} 6.2\,4 \\ + 8.4 \\ \hline \end{array}$

4. $\begin{array}{r} 5.2\,8 \\ + 2.0\,5 \\ \hline \end{array}$

Rewrite each number without an exponent.

5. $1^3 =$ _____

6. $10^2 =$ _____

7. $6^3 =$ _____

8. $7^2 =$ _____

Write in decimal notation.

9. $8 \times 10^3 + 4 \times 10^2 + 2 \times \dfrac{1}{10^1} =$ _____

10. $2 \times 10^2 + 6 \times 10^1 + 9 \times 10^0 + 5 \times \dfrac{1}{10^3} =$ _____

Fill in the missing numbers to make equivalent fractions.

11. $\dfrac{4}{5} = \dfrac{}{} = \dfrac{12}{} = \dfrac{}{20}$

12. $\dfrac{5}{9} = \dfrac{}{} = \dfrac{}{27} = \dfrac{}{}$

Add. Simplify your answer, if possible.

13. $\dfrac{1}{6} + \dfrac{2}{9} = $ ——

14. $\dfrac{1}{5} + \dfrac{7}{10} = $ ——

15. $\dfrac{1}{3} + \dfrac{3}{8} = $ ——

16. Fritha has \$4.75, and Rachel has \$6.29. Do they have enough money to buy a game that costs \$11.00?

17. Gary plans to buy a diamond necklace for his wife. The jeweler showed him the diamonds he wants to use for the necklace. They weighed 1.2 carats, 0.75 carats, and 1.15 carats. What was the total weight of the three diamonds?

18. Matthew must use 4/10 of his income for rent. How many fifths of Matthew's income is that?

19. Tess did 5/8 of the chores, and Dustin did 1/6 of them. What part of the chores have been done?

20. It rained for 3/5 of the days in April this year. Since there are 30 days in April, for how many days did it rain?

LESSON TEST

Add or subtract the decimal numbers.

1. $\begin{array}{r} 7.2\,5 \\ -\ 2.8 \\ \hline \end{array}$

2. $\begin{array}{r} 5.0\,7 \\ -\ 2.1\,1 \\ \hline \end{array}$

3. $\begin{array}{r} 6.2\,4\,9 \\ -\ 5.3\,7\,1 \\ \hline \end{array}$

4. $\begin{array}{r} 8.2 \\ +\ 0.9 \\ \hline \end{array}$

5. $\begin{array}{r} 6.1\,4 \\ +\ 0.3\,9\,5 \\ \hline \end{array}$

6. $\begin{array}{r} 1\,4.9 \\ +\ 0.1\,2\,4 \\ \hline \end{array}$

Add.

7. $\dfrac{2}{5} + \dfrac{1}{8} = \underline{}$

8. $\dfrac{3}{7} + \dfrac{2}{9} = \underline{}$

9. $\dfrac{2}{3} + \dfrac{1}{6} = \underline{}$

Write in decimal notation.

10. $2 \times 10^4 + 9 \times 10^3 + 1 \times \dfrac{1}{10^1} = \underline{}$

11. $5 \times 10^2 + 8 \times 10^1 + 3 \times \dfrac{1}{10^2} + 4 \times \dfrac{1}{10^3} = \underline{}$

Subtract.

12. $\dfrac{3}{4} - \dfrac{1}{5} =$ _____

13. $\dfrac{7}{8} - \dfrac{2}{3} =$ _____

14. $\dfrac{1}{4} - \dfrac{1}{6} =$ _____

15. Mom went to the store with a 20-dollar bill and spent $15.24 on groceries. How much money did she have left over?

16. Elizabeth planned to make bread to sell. When she bought the ingredients, she spent $7.35 for wheat berries, $2.35 for honey, $1.17 for oil, and $1.00 for yeast. How much did she spend in all?

17. Elizabeth sold the bread she made (see #16) for $35.00. What was her profit?

18. The triplets liked to talk on the phone. Pansy talked for 3.2 hours, Petunia talked for 2.3 hours, and Peony talked for 1.05 hours. How many hours did they spend talking in all?

19. Don has 2/10 of a dollar. How many cents does he have?

20. Three fourths of a cake was left after dinner. Later Brian ate one tenth of a whole cake. What part of the whole cake is left now?

Fill in the spaces above the lines with the appropriate words.

1. _____   _____   _____   _____
   1,000 grams (g)   100 grams (g)   10 grams (g)   1 gram (g)

2. _____   _____   _____   _____
   1,000 liters (L)   100 liters (L)   10 liters (L)   1 liter (L)

3. _____   _____   _____   _____
   1,000 meters (m)   100 meters (m)   10 meters (m)   1 meter (m)

Match the base unit with what it is used to measure.

4. volume or capacity   a. gram

5. mass or weight   b. meter

6. distance or length   c. liter

Add or subtract.

7. $\begin{array}{r} 1.2 \\ -\ 0.4 \\ \hline \end{array}$ 8. $\begin{array}{r} 5.0 \\ -\ 2.8 \\ \hline \end{array}$

9. $\begin{array}{r} 2\ 5.3\ 2 \\ +\ 1.0\ 6 \\ \hline \end{array}$ 10. $\begin{array}{r} 7.4\ 2 \\ -\ 2.3\ 9 \\ \hline \end{array}$

Add or subtract. Simplify when possible.

11. $\dfrac{1}{5} + \dfrac{1}{9} =$ ——

12. $\dfrac{3}{8} - \dfrac{1}{4} =$ ——

13. $\dfrac{7}{9} - \dfrac{2}{7} =$ ——

Change mixed numbers to improper fractions.

14. $5\dfrac{2}{3} =$ ——

15. $2\dfrac{6}{7} =$ ——

16. $1\dfrac{5}{8} =$ ——

17. Would the distance between two cities be measured in kilograms or kilometers?

18. April received a package with a mass of one kilogram. How many grams is that?

19. Patrick finished the race in 5.7 minutes, while Vaughn took 6.1 minutes. What is the difference in their times?

20. Write in exponential notation: 3,400.12

Fill in the spaces above the lines with the appropriate words.

1. _____ _____ _____ _____
   1 gram (g) $\frac{1}{10}$ gram (g) $\frac{1}{100}$ gram (g) $\frac{1}{1,000}$ gram (g)

2. _____ _____ _____ _____
   1 liter (L) $\frac{1}{10}$ liter (L) $\frac{1}{100}$ liter (L) $\frac{1}{1,000}$ liter (L)

3. _____ _____ _____ _____
   1 meter (m) $\frac{1}{10}$ meter (m) $\frac{1}{100}$ meter (m) $\frac{1}{1,000}$ meter (m)

Fill in the blank with the correct prefix.

4. The prefix for 1,000 is _____ .

5. The prefix for 100 is _____ .

6. The prefix for 10 is _____ .

Find the fraction of the number.

7. $\frac{1}{2}$ of 40 = _____

8. $\frac{3}{8}$ of 64 = _____

9. $\frac{2}{7}$ of 14 = _____

Change the improper fractions to mixed numbers.

10. $\frac{17}{8}$ = _____

11. $\frac{35}{4}$ = _____

12. $\dfrac{29}{3}$ = _____

13. Stephen is one yard tall, and Pippin is one meter tall. Who is taller?

14. Carmen bought five liters of juice. Is that closer to five quarts or five cups?

15. Would Jenna be more likely to use grams or kilograms to describe the size of her horse?

16. Which is greater: a kiloliter or a milliliter?

17. Is a kilometer closer to a mile or one half of a mile?

18. Is a kilogram closer to two pounds or one half of a pound?

19. During a heat wave, one stalk of corn grew 1.2 inches the first day, 2.14 inches the second day, and 1.75 inches the third day. How many inches did the corn grow in those three days?

20. After its growth spurt, the cornstalk (see #19) was 50.9 inches tall. How tall was it at the beginning of the three days?

LESSON TEST

8

Use the chart to help you convert the measures.

1 **kilo**unit	1 **hecto**unit	1 **deka**unit	1 unit	10 **deci**units	100 **centi**units	1,000 **milli**units
1,000 units	100 units	10 units	1 unit	1 unit	1 unit	1 unit

1. 45 kl = _____ liter

2. 6 g = _____ mg

3. 13 dm = _____ mm

Match the metric base unit with the unit of U.S. customary measure to which it is closest.

4. meter   a. 0.4 of an inch

5. centimeter   b. 1 yard

6. kilometer   c. 0.6 of a mile

7. liter   d. 2.2 pounds

8. gram   e. 1 quart

9. kilogram   f. 1/500 of a pound

Match the metric base unit with what it is used to measure.

10. liter   a. mass

11. gram   b. length or distance

12. meter   c. volume or capacity

ZETA LESSON TEST 8   17

Add.

13. $6\frac{3}{4}$

     $+\ 3\ \frac{1}{5}$
     _____

14. $11\frac{1}{9}$

     $+\ 4\frac{2}{7}$
     _____

15. $21\frac{2}{5}$

     $+\ 7\frac{5}{6}$
     _____

16.   Which is longer: 1,000 millimeters or 100 centimeters?

17.   Mindy compares two packages of strawberries that are the same price. For the better bargain, should she buy the package labeled one pound or the package labeled one kilogram?

18.   Jon picked up five small paper clips. About how much is their mass in grams?

19.   What is the mass of the five paper clips in #18 in centigrams?

20.   Marion and Anne each brought 4½ pies to the family reunion. How many pies did they bring together?

UNIT TEST Lessons 1-8

Rewrite each number without an exponent.

1. $1^6 =$ _____

2. $10^3 =$ _____

3. $9^2 =$ _____

Write in expanded notation.

4. $35.24 =$ _____

Write in exponential notation.

5. $90,000.16 =$ _____

Write in decimal notation.

6. $4 \times 10^3 + 1 \times 10^2 + 6 \times 10^1 + 8 \times 10^0 + 3 \times \dfrac{1}{10^1} + 2 \times \dfrac{1}{10^2} = 1 \times \dfrac{1}{10^3} =$

_____

Add or subtract the decimal numbers.

7.
$$
\begin{array}{r}
6.3\ 9 \\
-\ 3.5\ 4 \\
\hline
\end{array}
$$

8.
$$
\begin{array}{r}
5.0 \\
-\ 2.1\ 5 \\
\hline
\end{array}
$$

9.
$$
\begin{array}{r}
1.8\ 0\ 1 \\
-\ 0.9\ 9\ 9 \\
\hline
\end{array}
$$

10.
$$
\begin{array}{r}
8.3 \\
+\ 0.4 \\
\hline
\end{array}
$$

11.
$$1.23$$
$$+\ 0.147$$

12.
$$91.2$$
$$+\ \ 0.608$$

Find the fraction of the number.

13. $\frac{7}{8}$ of 96 = ____

14. $\frac{5}{9}$ of 81 = ____

Fill in the missing numbers to make equivalent fractions.

15. $\frac{6}{11} = \frac{}{} = \frac{}{33} = \frac{24}{}$

16. $\frac{9}{10} = \frac{18}{} = \frac{}{} = \frac{}{}$

Subtract. Simplify your answer, if possible.

17. $\frac{1}{6} - \frac{1}{7} =$ ____

18. $\frac{3}{4} - \frac{2}{5} =$ ____

19. $\frac{7}{9} - \frac{4}{9} =$ ____

Add and simplify as needed.

20.
$$1\frac{2}{3}$$
$$+\ 4\frac{1}{6}$$

21.
$$10\frac{3}{9}$$
$$+\ 7\frac{1}{4}$$

22.
$$34\frac{5}{6}$$
$$+\ 6\frac{1}{8}$$

Use the chart to help you convert using whichever method you prefer.

1 **kilo**unit	1 **hecto**unit	1 **deka**unit	1 unit	10 **deci**units	100 **centi**units	1,000 **milli**units
1,000 units	100 units	10 units	1 unit	1 unit	1 unit	1 unit

23. 12 m = _____ cm

24. 9 kg = _____ cg

25. 22 cl = _____ ml

Fill in the blanks with the appropriate metric unit.

26. A paper clip has a mass of about one _____ .

27. A _____ is a little more than two pounds.

28. The metric unit closest to a yard is a _____ .

29. The first part of the trip was 52.5 miles, and the second part was 15.06 miles. What was the total distance?

30. Jan went into a store with $50 and came out with $16.95. How much of her money had she spent?

LESSON TEST

Multiply.

1.   1.3
    × 0.2

2.   2.3
    × 1.2

3.   0.7
    × 0.1

4.   1.1
    × 0.6

5.   1.4
    × 1.2

6.   2.0
    × 0.4

Fill in the blanks. Write out your own chart if you need one.

7. 25 kg = _____ hg

8. 8 m = _____ cm

9. 31 cl = _____ ml

Fill in the blanks.

10. The metric unit closest to a quart is a _____.

11. The metric unit closest to one half of a mile is a _____.

12. The abbreviation kg stands for _____.

Subtract.

13. $12\frac{1}{4}$
   $-\ 3\frac{1}{5}$
   _____

14. 9
   $-\ 2\frac{7}{8}$
   _____

15. $21\frac{1}{3}$
   $-\ 6\frac{3}{4}$
   _____

16. One milliliter of water has a mass of about one gram. What is the mass of one centiliter of water?

17. Shawn planned to run 4.2 miles every day, but yesterday he only had time to run 0.2 of that distance. How far did Shawn run yesterday?

18. June bought three packs of chicken to cook at her barbecue. The labels on the packages gave the following weights: 3.3 pounds, 2.75 pounds, and 4.09 pounds. How many pounds of chicken did she buy in all?

19. June made four pies for dessert. After her guests went home, she had $1\frac{5}{6}$ pies left over. How many pies had been eaten?

20. The distance between two towns is 25.6 miles. Up to this point, 19.8 miles of the road have been repaired. How many miles are left to repair?

LESSON TEST

Multiply using whatever method you prefer.

1. $\begin{array}{r} 6.24 \\ \times 0.4 \\ \hline \end{array}$

2. $\begin{array}{r} 4.4 \\ \times 1.7 \\ \hline \end{array}$

3. $\begin{array}{r} 0.67 \\ \times 0.12 \\ \hline \end{array}$

4. $\begin{array}{r} 7.18 \\ \times 0.7 \\ \hline \end{array}$

5. $\begin{array}{r} 3.3 \\ \times 1.5 \\ \hline \end{array}$

6. $\begin{array}{r} 0.85 \\ \times 0.01 \\ \hline \end{array}$

Fill in the blanks.

7. 9 kg = _____ cg

8. 48 m = _____ mm

Add or subtract.

9. 12.17 + 147.09 = _____

10. 5.13 + 0.26 = _____

11. 18.7 − 4.8 = _____

12. $4\frac{3}{4} + 1\frac{1}{10} =$ _____

13. $13\frac{1}{2} - 4\frac{1}{8} =$ _____

14. $20 - 10\frac{1}{5} =$ _____

Multiply.

15. $\dfrac{5}{6} \times \dfrac{1}{4} =$ ___

16. $\dfrac{3}{8} \times \dfrac{2}{3} =$ ___

17. $\dfrac{5}{6} \times \dfrac{2}{3} =$ ___

18. Kimberly bought four gardening books for $9.24 each. How much did the four books cost in all?

19. Joel drove 642 kilometers yesterday. A kilometer is about 0.6 of a mile. Approximately how many miles did Joel drive?

20. Carl plans to order mulch for his garden. If the cost is $59.50 for a pickup truck load, how much will he have to pay for 1.75 loads?

LESSON TEST

Write each percentage as a decimal.

1. 35% = _____

2. 6% = _____

3. 19% = _____

4. 15% = _____

5. 58% = _____

6. 7% = _____

Write each percentage as a fraction. Simplify as needed.

7. 2% = ——— = ———

8. 23% = ——— = ———

9. 44% = ——— = ———

Write each fraction as a decimal and as a percentage.

10. $\frac{1}{5}$ = _____ = _____

11. $\frac{1}{2}$ = _____ = _____

12. $\frac{1}{4}$ = _____ = _____

13. $\frac{3}{4}$ = _____ = _____

Multiply.

14.
$$9.3 \times 0.01$$

15.
$$3.5 \times 4.5$$

16.
$$2.56 \times 0.81$$

17. $4\frac{2}{7} \times 1\frac{1}{2} =$

18. $1\frac{2}{3} \times 6\frac{4}{5} =$

19. Kyle bought a meal that cost $15.96 and left a 15% tip. What was the total cost of the meal and the tip? (Round your answer to the nearest cent.)

20. A kilometer is about 60% of a mile. When Gerry finished a 20-kilometer bike race, about how many miles had he ridden?

LESSON TEST

Change each mixed number to a percentage and a decimal.

1. $3\dfrac{3}{4} = \dfrac{}{100} + \dfrac{}{100} = \dfrac{}{100} = \underline{} = \underline{}$

2. $1\dfrac{2}{5} = \dfrac{}{100} + \dfrac{}{100} = \dfrac{}{100} = \underline{} = \underline{}$

3. $6\dfrac{9}{10} = \dfrac{}{100} + \dfrac{}{100} = \dfrac{}{100} = \underline{} = \underline{}$

4. $2\dfrac{1}{2} = \dfrac{}{100} + \dfrac{}{100} = \dfrac{}{100} = \underline{} = \underline{}$

Write each percentage as a decimal and as a simplified fraction.

5. 35% = _____ = _____

6. 11% = _____ = _____

Fill in the blanks with the correct U.S. customary unit of measure.

7. One meter is a little longer than a _____ .

8. 28 g ≈ 1 _____

9. 2.54 cm ≈ 1 _____

10. 1 cm ≈ 0.4 _____

11. 1 kg ≈ 2.2 _____

12. 1 km ≈ 0.6 _____

13. 1 liter ≈ 1.06 _____

Divide. Use the Rule of Four to make the denominators the same.

14. $\dfrac{3}{5} \div \dfrac{1}{3} =$ ____

15. $\dfrac{3}{4} \div \dfrac{7}{8} =$ ____

16. $\dfrac{2}{5} \div \dfrac{3}{8} =$ ____

17. Kevin ordered a CD that cost $11.25. The tax was 4%, and the shipping was 6%. Use a percent greater than 100 to find his total bill.

18. Ken has collected 25 baseball cards. His goal is to have 400% of that number. How many cards does he hope to collect in all?

19. Virginia has 1/2 of a cake left over. She wants to serve each of her guests 1/10 of a whole cake. How many guests can she serve?

20. A $75 dress is on sale at 25% off. How much money will Pam save if she buys the dress on sale?

Use the given information to shade or color the chart. Shade or color the boxes in the key to match.

1. Thirty-five percent of the music was performed by the orchestra alone. Five percent of the music was performed by the chorus alone. Sixty percent of the music was performed by both chorus and orchestra. (Each section of the chart is 5%.)

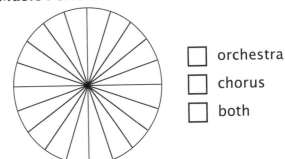

Music Performed at the Concert

☐ orchestra
☐ chorus
☐ both

2. The concert lasted two hours. For how many minutes did the chorus perform alone?

3. What kind of weather occurred most often in April?

4. Since there are 30 days in April, how many were rainy?

Weather in April

■ sunny
▨ rainy
▧ cloudy

40%   30%   30%

5. How many days in April were cloudy?

Write the mixed number as a percentage and as a decimal.

6. $4\frac{3}{4} = \frac{}{100} + \frac{}{100} = \frac{}{100} = \underline{\hspace{2cm}} = \underline{\hspace{2cm}}$

Write each percentage as a decimal.

7. 25% = _____

8. 16% = _____

9. 6% = _____

10. 125% = _____

Solve.

11. $3.61 \times 0.07 =$ _____

12. $0.9 - 0.14 =$ _____

13. $2.69 + 8.7 =$ _____

Divide.

14. $8\frac{4}{5} \div 4\frac{2}{3} =$

15. $3\frac{2}{5} \div 6\frac{5}{7} =$

16. $14\frac{1}{2} \div 5\frac{2}{3} =$

17. Dustin's favorite snack used to cost $1.45 a box. If the cost has gone up 15%, what is the price now?

18. Madelyn bought a 10-kilogram bag of potatoes. How many grams is that?

19. A craftsman can repair about 1.5 antique books a week. How many books can he repair in six weeks?

20. One gallon equals four quarts. Approximately how many liters is that?

LESSON TEST

Estimate and then multiply using whichever method you prefer.

1. $\begin{array}{r} 0.1\,2\,3 \\ \times\,0.4\,5\,6 \\ \hline \end{array}$

2. $\begin{array}{r} 3.0\,5\,2 \\ \times\,6.1\,9\,3 \\ \hline \end{array}$

3. $\begin{array}{r} 0.2\,8\,1 \\ \times\,0.2\,6\,9 \\ \hline \end{array}$

4. $\begin{array}{r} 1.5\,0\,4 \\ \times\,5.1\,6\,7 \\ \hline \end{array}$

Write each percentage as a simplified fraction or whole number.

5. $500\% = \underline{\quad}$

6. $50\% = \underline{\quad}$

7. $38\% = \underline{\quad}$

8. $2\% = \underline{\quad}$

Write the mixed number as a percentage and as a decimal.

9. $9\dfrac{2}{5} = \dfrac{\quad}{100} + \dfrac{\quad}{100} = \dfrac{\quad}{100} = \underline{\quad} = \underline{\quad}$

Fill in the blanks.

10. $16\ m = \underline{\quad}\ mm$

11. $35\ kl = \underline{\quad}\ hl$

12. $1\ g = \underline{\quad}\ dg$

Divide using the shortcut.

13. $\dfrac{5}{8} \div \dfrac{2}{3} =$

14. $\dfrac{1}{5} \div \dfrac{3}{8} =$

15. $\dfrac{1}{2} \div \dfrac{3}{5} =$

16. Vicki bought 10.4 gallons of fuel at $1.639 a gallon. What was the total cost?

17. One meter equals about 39.37 inches. About how many inches are in 3.2 meters?

18. Nathan started working at $6 an hour. He did such a good job that he got a 20% raise. What is his new hourly rate?

19. Julianne ordered catalog items that cost $25.60, $11.19, and $45.21. What was the total cost of the three items?

20. The shipping for Julianne's order (see #19) was 8%, and the tax was 3%. What was her total cost?

LESSON TEST

Convert using whichever method you prefer.

1. _____ m = 7 cm

2. _____ kl = 10 dal

3. _____ g = 500 mg

4. _____ hm = 25 m

5. 8 cl = _____ ml

6. 95 g = _____ cg

Estimate and then multiply using whichever method you prefer.

7. $\begin{array}{r} 1.4\,6\,5 \\ \times\,0.0\,0\,7 \\ \hline \end{array}$

8. $\begin{array}{r} 2\,3\,9.0\,1\,6 \\ \times\ \ 0.1\,3\,4 \\ \hline \end{array}$

Write each percentage as a decimal or a whole number.

9. 6% = _____

10. 45% = _____

11. 130% = _____

12. 200% = _____

Find the fraction of the number.

13. $\frac{1}{2}$ of 34 = _____

14. $\frac{2}{9}$ of 36 = _____

15. $\frac{3}{5}$ of 100 = _____

Divide using the reciprocal.

16. $4\frac{1}{8} \div 1\frac{1}{5} =$

17. $8\frac{2}{3} \div \frac{1}{6} =$

18. What is a mass of 1,000,000 milligrams expressed in kilograms?

19. The path to Cameron's back door is 50 dekameters long. How many millimeters long is the path?

20. A farmer's corn crop was 108% of what he expected. If he expected to harvest 35 tons, what was the actual harvest?

LESSON TEST

Use a decimal approximation of π to find the area and the circumference of each circle.

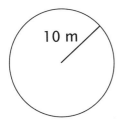

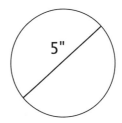

1. Area ≈ _____

2. Circumference ≈ _____

3. Area ≈ _____

4. Circumference ≈ _____

Convert using whichever method you prefer.

5. _____ m = 61 mm

6. _____ kg = 400 cg

7. 13 cl = _____ ml

Multiply.

8. $0.07 \times 0.04 =$ ____

9. $72.3 \times 0.9 =$ ____

10. $204 \times 0.11 =$ ____

Divide using the reciprocal.

11. $8\frac{1}{2} \div 2\frac{1}{4} =$

12. $3\frac{1}{6} \div 1\frac{2}{3} =$

13. $\frac{9}{10} \div \frac{2}{5} =$

Find the perimeter of each rectangle.

14. P = _____

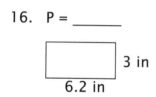

4 m

9 m

15. P = _____

4.5 ft

4.5 ft

16. P = _____

3 in

6.2 in

17. The Nelsons built a round swimming pool with a radius of six feet. Approximately how much area did the pool cover?

18. Sandy is sewing fringe to the edge of a round rug for her living room. If the rug measures five feet across, how many feet of fringe does she need?

19. How many yards of fence are needed for a square play yard that measures eight yards on a side?

20. Louise walked two hectometers to work every morning. How many meters did she walk each morning?

Multiply.

1.   8.6 1
   × 0.7

2.   1.4
   × 2.6

3.   0.1 8
   × 0.9 5

4.   0.1 0 6
   × 0.3 5 2

5.   2.6 0 5
   × 6.2 7 9

6.   3.1 9 1
   × 4.2 6 0

Write each percentage as a decimal or a whole number.

7. 1% = _____

8. 10% = _____

9. 100% = _____

Write each percentage as a simplified fraction.

10. 5% = —— = ——

11. 48% = —— = ——

12. 16% = —— = ——

Write each fraction as a decimal and as a percentage.

13. $\dfrac{1}{4}$ = _____ = _____

14. $\dfrac{2}{5}$ = _____ = _____

15. $\dfrac{1}{2}$ = _____ = _____

16. $\dfrac{3}{4}$ = _____ = _____

Write the mixed number as a percentage and as a decimal.

17. $6\dfrac{1}{4}$ = $\dfrac{}{100}$ + $\dfrac{}{100}$ = $\dfrac{}{100}$ = _____ = _____

Convert using whichever method you prefer.

18. _____ m = 200 mm

19. 81 kg = _____ cg

Use the given information to shade or color the chart. Shade or color the boxes in the key to match.

20. 5% went for electricity to make ice cubes.

80% went for the ingredients for the lemonade.

15% went for paper cups.

(Each section of the chart is 5%.)

21. If total expenses were $35, how much was spent on ingredients?

Lemonade Stand Expenses

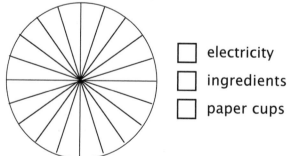

☐ electricity
☐ ingredients
☐ paper cups

Use a decimal approximation of π to find the area and the circumference of the circle.

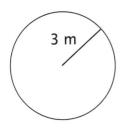

22. Area ≈ _____

23. Circumference ≈ _____

Multiply.

24. $5\frac{2}{5} \times 4\frac{1}{5} =$

25. $2\frac{1}{3} \times \frac{5}{9} =$

Divide.

26. $7\frac{1}{2} \div 1\frac{1}{4} =$

27. $6\frac{2}{3} \div \frac{2}{5} =$

28. Levi got 88% of the questions correct on his history test. If there were 50 questions on the test, how many were correct?

29. Zac ordered a part that cost $135 for his car. He had to pay 6% tax and 10% shipping. Use a percent greater than 100 to find his total cost.

30. What is the perimeter of a rectangle that measures 10.16 inches by 9.25 inches? Give the answer without rounding.

42

ZETA

UNIT TEST II

LESSON TEST

Divide the decimals and check by multiplying.

1. $2 \overline{)\,0.0\,0\,4}$

2. Check for #1

3. $8 \overline{)\,1.7\,6}$

4. Check for #3

5. $4 \overline{)\,0.2\,8}$

6. Check for #5

7. $5 \overline{)\,1.0}$

8. Check for #7

9. $7 \overline{)\,3.5}$

10. Check for #9

11. $3 \overline{)\,0.0\,0\,9}$

12. Check for #11

The circle has a diameter of 2.4 miles. Answer the questions.

13. What is the approximate area of the circle? _____

14. What is the approximate circumference of the circle? _____

Find the perimeter of each parallelogram.

15. P = _____

30 in

40 in

16. P = _____

3.2 m

7.9 m

17. P = _____

4¾ in

6¼ in

18. Mr. Anderson had $40.95 that he wanted to divide evenly among his seven children. How much should each child receive?

19. Dianne has 71.6 miles to bike in the next four hours. If she divides the distance evenly, how far should she bike in each hour?

20. Joy worked 3.5 hours, 6.9 hours, and 4.3 hours in the last three days. If she earns $7.10 an hour, how much did she earn in the last three days?

LESSON TEST

Divide the decimals and check by multiplying.

1. $0.004 \overline{)484}$

2. Check for #1

3. $0.18 \overline{)36.18}$

4. Check for #3

5. $6 \overline{)22.2}$

6. Check for #5

7. $9 \overline{)0.0054}$

8. Check for #7

A circle has a radius of 3.5 inches. Answer the questions about the circle.

9. What is the approximate area of the circle? _____

10. What is the approximate circumference of the circle? _____

Simplify each expression.

11. $13^2 =$ ____

12. $6^2 =$ ____

13. $10^2 = $ ____

Find the perimeter of each triangle.

14. P = _____

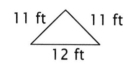

15. P = _____

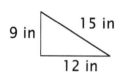

16. P = _____

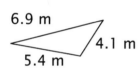

17. Paul wants to change $19 into quarters. Divide by 0.25 to find how many quarters he would receive.

18. There are 5.4 pounds of raisins in the bag. If Mom uses 0.45 of a pound in each batch of cookies, how many batches can she make?

19. What is the perimeter of a square that measures 13.9 meters on a side?

20. Gena is 3,160 grams heavier than she was last year at this time. How many kilograms is that?

LESSON TEST

Solve for the unknown. Check your answers.

1. $0.19G = 38$

2. Check for #1

3. $0.008Y = 2$

4. Check for #3

5. $0.6G = 24$

6. Check for #5

7. $0.11Y = 55$

8. Check for #7

Divide and check by multiplying.

9. $0.006 \overline{\smash{\big)}\ 6.018}$

10. Check for #9

11. $15 \overline{\smash{\big)}\ 3.0}$

12. Check for #11

Find the area of each rectangle or square.

13. A = _____

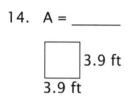
4 m

8 m

14. A = _____

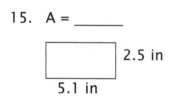
3.9 ft

3.9 ft

15. A = _____

2.5 in

5.1 in

16. Forty-five percent of the job is done. If 90 hours have been worked so far, what is the total time needed for the job? Write an equation and solve.

17. How many hours are left to finish the job in #16?

18. Toni has wrapped 0.75 of her gifts. If 12 gifts have been wrapped, what was the total number she had to wrap? Write an equation and solve.

19. A check for $308 is to be divided evenly among eight people. How much money will each person receive?

20. Which has more mass: a 5,000-gram package or a 50-kilogram package?

LESSON TEST

Divide the decimals and check by multiplying.

1. $0.5 \overline{)27.5}$

2. Check for #1

3. $0.32 \overline{)6.8}$

4. Check for #3

5. $0.04 \overline{)27.2}$

6. Check for #5

7. $0.28 \overline{)4.2}$

8. Check for #7

9. $35 \overline{)1.96}$

10. Check for #9

11. $0.002 \overline{)8.54}$

12. Check for #11

Solve for the unknown. Divide to the hundredths place. Include a fraction in your answer if there is still a remainder.

13. $0.12Q = 0.36$

14. $2.5W = 0.75$

Solve.

15. $5\dfrac{2}{3} \div \dfrac{6}{7} =$

16. $24 - 12\dfrac{1}{10} =$

17. $\dfrac{4}{5} \times \dfrac{10}{11} =$

Find the area of the parallelogram.

18. A = _____

19. Paul walked 27.3 miles. He stopped to rest every 9.1 miles. Into how many parts did he divide his walk?

20. Debra has $66.36. How many items can she buy that cost $3.16 apiece?

LESSON TEST

Divide to the thousandths place and round to the nearest hundredth.

1. $0.4\overline{)8.5\ 3}$

2. $16\overline{)3.9\ 3}$

3. $0.0\ 9\overline{)0.1\ 8\ 6}$

Divide until you see a pattern. Write the answer with a line over the repeating digits.

4. $1.1\overline{)8.2}$

5. $0.2\ 2\overline{)9.7}$

6. $6\overline{)7}$

Solve for the unknown. Divide to the hundredths place. Include a fraction in your answer if there is still a remainder.

7. $30X = 0.8$

8. $0.5Y = 0.123$

9. $2.3F = 6.41$

Solve.

10. 65 – 0.45 = _____

11. 7.03 + 0.2 = _____

12. 200 – 0.02 = _____

Find the area of each triangle.

13. A = _____

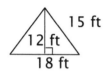

15 ft
12 ft
18 ft

14. A = _____

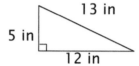

13 in
5 in
12 in

15. A = _____

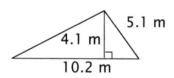

5.1 m
4.1 m
10.2 m

16. The area of a rectangle is 51.84 square meters. If the base of the rectangle is 5.4 meters long, what is the height of the rectangle?

17. Dennis spent $25.45 for a meal and left a 20% tip. What was the total amount that he spent?

18. Justine cut out paper circles with a diameter of 16 inches. What was the area of each circle?

19. Is your breakfast cereal more likely to be measured in grams or liters?

20. Give the difference in grams between an object with a mass of 31 kilograms and an object with a mass of 310 grams.

LESSON TEST

Solve for the unknown. Check your answer by using it in the original problem.

1. 0.09G + 0.38 = 0.425

2. Check for #1

3. 0.4X + 3.6 = 6.4

4. Check for #3

5. 1.5X + 4 = 4.165

6. Check for #5

Divide to the thousandths place and round to the nearest hundredth.

7. $0.9 \overline{)0.8\ 5}$

8. $4 \overline{)1\ 6.1\ 3}$

9. $0.0\ 7 \overline{)0.6\ 2\ 8}$

Convert using whichever method you prefer.

10. _____ m = 1,800 mm

11. _____ kl = 549 L

12. 4.5 cg = _____ mg

Find the volume of each cube.

13. V = _____

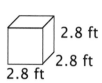

14. V = _____

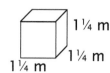

Find the area and perimeter of each figure.

15. A = _____

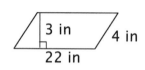

16. P = _____

17. A = _____

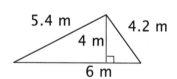

18. P = _____

19. Natalie bought six ice cream cones at $1.69 each. She had to pay a tax of 3%. What was the total cost of her treats (to the nearest cent)?

20. Crystal bought a dress for 45% of the original cost. After adding $1.35 for taxes, she paid a total of $23.85. What was the original cost of the dress? Write an equation and solve.

LESSON TEST

Convert each fraction to a decimal and then to a percentage. Include a fraction if needed. Do not round.

1. $\dfrac{3}{10}$ = _____ = _____%

2. $\dfrac{1}{7}$ = _____ = _____%

3. $\dfrac{2}{3}$ = _____ = _____%

4. $\dfrac{5}{6}$ = _____ = _____%

5. $\dfrac{1}{13}$ = _____ = _____%

6. $\dfrac{8}{9}$ = _____ = _____%

Solve for the unknown. Check your answer by using it in the original problem.

7. $0.9R + 0.5 = 0.77$

8. Check for #5

Divide until you see a pattern. Write the answer with a line over the repeating digits.

9. $6\,\overline{)\,3\,1.7}$

10. $0.3\,\overline{)\,9\,1}$

11. $0.0\,9\,\overline{)\,0.4\,2\,5}$

Find the volume of each rectangular solid.

12. V = _____

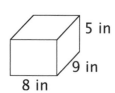

    5 in

    9 in

8 in

13. V = _____

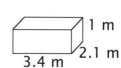

    1 m

    2.1 m

3.4 m

14. V = _____

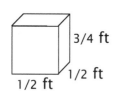

    3/4 ft

    1/2 ft

1/2 ft

15. On game night, Tanny answered 35 out of 50 questions correctly in a trivia game. What was the percentage of questions she answered correctly?

16. Dean ate 1/8 of his birthday cake. What percent of the cake did he eat?

17. Fresh salmon is selling for $5.65 a pound. Use a decimal to find how much 3/5 of a pound would cost.

18. Joseph makes wooden toys for sale. He hopes to finish making 45 toys by the end of the week. He has already finished 80% of that number. How many toys are still left to make?

19. The price of the meal was $11.20. The tip was 15% of the price, and the tax was 5% of the price. How much change should you receive from a 20-dollar bill?

20. The area of a square is 100 square feet. If one side is 10 feet long, how long are the other sides?

Divide and check by multiplying.

1. $4\overline{)0.0\,0\,8}$

2. Check for #1

3. $6\overline{)0.3}$

4. Check for #3

5. $0.2\overline{)8\,6}$

6. Check for #5

7. $0.0\,9\overline{)0.0\,0\,6\,3}$

8. Check for #7

9. $2.1\overline{)3\,3.6}$

10. Check for #9

11. $0.0\,0\,5\overline{)9.0\,5}$

12. Check for #11

Solve for the unknown. Check your answer by using it in the original problem.

13. $6.3W = 11.34$

14. Check for #13

15. $0.7X + 0.08 = 17.58$

16. Check for #15

Find the area and perimeter of each figure.

17. A = _____

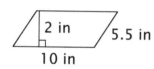

18. P = _____

19. A = _____

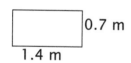

20. P = _____

21. A = _____

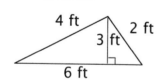

22. P = _____

Divide to the thousandths place and round to the nearest hundredth.

23. $6 \overline{)7\,3.4}$

Divide until you see a pattern. Write the answer with a line over the repeating digits.

24. $0.3 \overline{)4.5\,5}$

Divide to the hundredths place. Include a fraction in your answer if there is still a remainder.

25. $8 \overline{)7.0}$

Change each fraction to a decimal and then to a percentage. Include a fraction after the percent if needed. Do not round.

26. $\dfrac{9}{10}$ = _____ = _____%

27. $\dfrac{2}{7}$ = _____ = _____%

28. $\dfrac{1}{3}$ = _____ = _____%

29. $\dfrac{3}{4}$ = _____ = _____%

30. A rectangular solid measures 8 feet by 9 feet by 7.2 feet. What is its volume?

Change each decimal to a simplified fraction.

1. $0.85 = \underline{\quad} = \underline{\quad}$

2. $0.16 = \underline{\quad} = \underline{\quad}$

3. $0.2 = \underline{\quad} = \underline{\quad}$

4. $0.005 = \underline{\quad} = \underline{\quad}$

Change each fraction to a decimal and then to a percentage. Include a fraction in the hundredths place if needed. Do not round.

5. $\dfrac{2}{9} = \underline{\quad} = \underline{\quad}\%$

6. $\dfrac{4}{5} = \underline{\quad} = \underline{\quad}\%$

7. $\dfrac{1}{2} = \underline{\quad} = \underline{\quad}\%$

8. $\dfrac{2}{3} = \underline{\quad} = \underline{\quad}\%$

Solve for the unknown. Check your answer by using it in the original problem.

9. $0.75X = 7.5$

10. Check for #9

11. $1.4X + 5 = 6.82$

12. Check for #11

Divide to the hundredths place. Include a fraction in your answer if there is still a remainder.

13. $3\overline{)40}$

14. $0.11\overline{)0.268}$

Find the average of each series of numbers.

15. 1, 5, 9   Average = _____

16. 2, 2, 5, 7   Average = _____

17. Zachary's weight is 130% of Noah's weight. If Noah weighs 75 pounds, what does Zachary weigh?

18. A rectangle has two sides that measure 45 inches and two sides that measure 31 inches. What is the perimeter of the rectangle?

19. A circular irrigation system has a radius of 60 feet. How much land can be watered with the system?

20. Beth walked 3.5 kilometers. How many meters did she walk?

LESSON TEST

Find the meAn, MeDian, and MOde for each list of data.

1. 8, 13, 17, 17, 19, 19, 19

    mean = _____ , median = _____ , mode = _____

2. 80, 80, 90, 100, 70

    mean = _____ , median = _____ , mode = _____

3. 8, 15, 10, 8, 9

    mean = _____ , median = _____ , mode = _____

4. 3, 7, 11, 6, 3

    mean = _____ , median = _____ , mode = _____

Change each decimal to a simplified fraction.

5. 0.53 = ——

6. 0.44 = —— = ——

Convert each fraction to a decimal and then to a percentage. Include a fraction if needed. Do not round.

7. $\frac{23}{40}$ = _____ = _____%

8. $\frac{4}{6}$ = _____ = _____%

Solve for the unknown. Check your answer by using it in the original problem.

9. $9X + 3.4 = 7.9$

10. Check for #9

Simplify each expression.

11. $10^2 =$ _____

12. $3^3 =$ _____

13. $1^6 =$ _____

Find the perimeter and area of the parallelogram and the square.

14. P = _____

15. A = _____

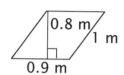

16. P = _____

17. A = _____

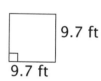

18. A teacher gave a quiz. Her students earned the following scores: 4, 3, 2, 3, 4, 2, 4, 6, 8. What was the average student score? What is another term for the average score?

19. Which score appeared most often in the list? What is that number called?

20. If the scores are arranged in order from least to greatest, which one will be in the middle? What is that number called?

LESSON TEST

There are 1,000 entries in a free drawing for a turkey.

1. If you enter 10 times, what is the probability of winning the turkey?

2. If you enter 300 times, what is the probability of winning?

3. If you enter 300 times, what is the probability of not winning?

Use the dominoes to answer the questions.

4. What is the probability of picking a domino with a 1 on it?

5. What is the probability of picking a domino with a 3 on it?

6. What is the probability of picking a domino with a 1 or a 2 on it?

Find the meAn, MeDian, and MOde for each list of data.

7. 9, 3, 7, 8, 8

    mean = _____ , median = _____ , mode = _____

8. 12, 9, 9, 17, 13

    mean = _____ , median = _____ , mode = _____

Change each decimal to a simplified fraction.

9. 0.5 = —— = ——

10. 0.92 = —— = ——

11. 0.09 = —— = ——

12. 0.825 = —— = ——

Change each fraction to a decimal and then to a percentage.

13. $\frac{1}{3}$ = _____ = _____%

14. $\frac{7}{8}$ = _____ = _____%

15. A skater spins in one spot with one leg extended. If the skater's leg is 3.2 feet long, about how far does her raised foot travel with each spin? (Round your answer to the nearest whole foot.)

16. Sixteen percent of the customers that visited the gift shop bought greeting cards. If 200 people visited the shop, how many bought greeting cards?

17. Which is the greater distance: 500 yards or 500 meters?

18. Mental Math! Eight plus four, minus six, plus three, times two, equals _____ .

LESSON TEST

Fill in the blanks.

1. The origin of a ray may also be called its _____.

2. A point has no _____ or _____.

3. A _____ has one endpoint.

4. A _____ has two endpoints.

5. A line has _____ but no _____.

Matching.

6. ↔   a. infinity

7. →   b. point

8. —   c. line

9. ·   d. ray

10. ∞   e. line segment

Use letters and symbols to name each figure.

11. _____   X •————————— Y •—————————→

12. _____   ←————— Q •—————————— R •—————→

Find the meAn, MeDian and MOde for the given data.

1, 2, 1, 5, 1

13. mean = _____

14. median = _____

15. mode = _____

Solve.

16. $2.05 \times 1.3 =$ _____   17. $1.43 \div 0.11 =$ _____

18. $1.75 + 2.25 =$ _____

19. Coins have two sides: heads and tails. If Tamara tosses a penny in the air, what is the probability that it will land heads up?

20. Wesley told his mother that 50% of his chores were done. What fractional part of the chores are done? If there were eight chores to start with, how many has Wesley finished?

Fill in the blanks.

1. A _____ is a figure with no length or width.

2. A _____ has infinite length and width and is two-dimensional.

3. The study of flat, two-dimensional shapes is _____ geometry.

4. Two numbers that have the same value are _____ .

5. Two shapes that are exactly the same are _____ .

6. Two objects that are the same shape but different sizes are _____ .

Write or draw the symbol for each word.

7. line

8. similar

9. infinity

10. ray

11. point

12. line segment

13. plane

14. congruent

Convert each fraction to a decimal and then to a percentage. Include a fraction if needed. Do not round.

15. $\dfrac{1}{2}$ = _____%

16. $\dfrac{1}{4}$ = _____%

17. $\dfrac{3}{4}$ = _____%

18. $\dfrac{1}{3}$ = _____%

19. $\dfrac{1}{5}$ = _____%

20. $\dfrac{2}{3}$ = _____%

LESSON TEST

Fill in the blanks.

1. A single angle can be shown by drawing two _____ with the same endpoint.

2. A right angle is _____ of a circle.

3. A right angle has a measure of _____ degrees.

4. There are _____ degrees in a circle.

5. The common endpoint of two rays that form an angle is called the _____.

6. Two objects that are the same shape but different sizes are _____.

7. Two figures that have exactly the same shape and size are _____.

8. A point has no _____ or _____.

9. Use the letters to name the given angle correctly..

   _____ or _____

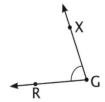

Write each decimal as a simplified fraction.

10. 0.5 = ——

11. 0.08 = ——

12. 0.565 = ——

Solve.

13. $\dfrac{4}{5} \times \dfrac{3}{10} =$ _____

14. $\dfrac{1}{8} + \dfrac{7}{12} =$ _____

15. $\dfrac{5}{8} - \dfrac{1}{2} =$ _____

16. A game has 150 cards, and five of them are bonus cards. If all the cards are on the table, what is the probability of getting a bonus card when choosing a card at random?

17. Given the following list of data, is 9 the mean, the median, or the mode?

    2, 7, 9, 8, 9, 4, 6

18. Four children have the following weights: 45.5 pounds, 61 pounds, 37.2 pounds, and 16.34 pounds. What is the average weight of the children?

LESSON TEST

Fill in the blanks.

1. The measure of a right angle is _____ .

2. An angle with a measure less than 90° but greater than 0° is a(n) _____ angle.

3. An angle with a measure of 180° is called a(n) _____ angle.

4. An angle with a measure less than 180° but greater than 90° is a(n) _____ angle.

Identify each kind of angle pictured.

5. _____

6. _____

7. _____

8. _____

Write each number as a percentage.

9. 4.8 = ⎯⎯⎯

10. 0.91 = ⎯⎯⎯

11. 0.09 = ⎯⎯⎯

Find the percentage of each number.

12. 25% of 300 = ⎯⎯⎯

13. 150% of 80 = ⎯⎯⎯

14. 30% of 75 = ⎯⎯⎯

Divide the fractions.

15. $\dfrac{2}{3} \div \dfrac{1}{8} =$ —

16. $\dfrac{4}{9} \div \dfrac{1}{3} =$ —

17. What number has the same value as 15^2?

18. Michael tied a weight to a 6.1-foot rope and swung it in a circle. About how far did the weight travel in one swing?

UNIT TEST Lessons 24-30

Change each decimal to a simplified fraction.

1. $0.52 =$ —— $=$ ——

2. $0.08 =$ —— $=$ ——

3. $0.3 =$ ——

4. $0.575 =$ —— $=$ ——

Find the mean, median, and mode for each list of data.

5. 10, 17, 12, 10, 11

    mean = _____ , median = _____ , mode = _____

6. 8, 13, 17, 17, 19, 19, 19

    mean = _____ , median = _____ , mode = _____

A book has 280 pages. Of the 280 pages, 210 have printing, 40 have pictures, and 30 are blank.

7. What is the probability of opening to a picture?

8. What is the probability of opening to a printed page?

9. What is the probability of **not** opening to a printed page?

Use letters and symbols to name the figures shown below.

10. _____

11. _____

12. _____

13. _____

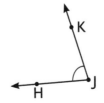

ZETA UNIT TEST IV

75

Fill in the blanks.

14. A _____ has one endpoint.

15. A _____ has two endpoints.

16. A line has _____ but no _____.

17. A _____ has no length or width.

18. A _____ has infinite length and width and is said to be two-dimensional.

19. Two shapes that are exactly the same are _____.

20. Two numbers that have the same value are _____.

21. Two figures that are the same shape but different sizes are _____.

22. The common endpoint of two rays that form an angle is called the _____.

23. There are _____ degrees in a circle.

24. The measure of a right angle is _____.

25. An angle that has a measure less than 90° but greater than 0° is a(n) _____ angle.

26. An angle with a measure of 180° is called a(n) _____ angle.

27. An angle with a measure less than 180° but greater than 90° is a(n) _____ angle.

Match each word with its symbol.

28. angle   a. ~

29. plane   b. →

30. infinity   c. ∠

31. ray   d. ↔

32. line   e. ▱

33. congruent   f. ∞

34. similar   g. ≅

FINAL TEST Lessons 1-30

Rewrite each number without an exponent.

1. $1^6 =$ _____

2. $8^2 =$ _____

3. $10^3 =$ _____

Write in decimal notation.

4. $5 \times 10^3 + 2 \times 10^2 + 7 \times 10^1 + 1 \times 10^0 + 3 \times \dfrac{1}{10^1} + 4 \times \dfrac{1}{10^2} + 9 \times \dfrac{1}{10^3} =$

_____

Add or subtract the decimal numbers.

5.   7.5 2
     −1.8 5

6.   6.0
     + 5.2 8

7.   3 2.0 4 1
     − 0.5 9 6

Multiply the decimal numbers.

8.   2.4 9
     × 0.6

9.   1.7
     × 3

10.   0.0 0 4
     × 0.0 5

Convert using whatever method you prefer.

11. 13 km = _____ cm

12. _____ g = 250 mg

Write each percentage as a decimal.

13. 5% = _____

14. 65% = _____

Write each percentage as a simplified fraction.

15. 25% = ⎯⎯⎯ = ⎯⎯⎯

16. 32% = ⎯⎯⎯ = ⎯⎯⎯

Convert each fraction to a decimal and then to a percentage. Include a fraction if needed. Do not round.

17. $\dfrac{8}{10}$ = _____ = _____%

18. $\dfrac{5}{6}$ = _____ = _____%

Write the mixed number as a percentage and as a decimal.

19. $4\dfrac{3}{5}$ = $\dfrac{}{100}$ + $\dfrac{}{100}$ = $\dfrac{}{100}$ = _____ = _____

Change each decimal to a simplified fraction.

20. 0.78 = ⎯⎯⎯ = ⎯⎯⎯

21. 0.03 = ⎯⎯⎯ = ⎯⎯⎯

Divide to the thousandths place. Round to the nearest hundredth.

22. 4 $\overline{)\,1\,3.3}$

23. 7 $\overline{)\,4.5\,8}$

Divide until you see a pattern. Write the answer with a line over the repeating digits.

24. 0.6 $\overline{)\,3\,9.4}$

25. 0.0 3 $\overline{)\,0.0\,2\,2}$

Divide to the hundredths place. Include a fraction in your answer if there is still a remainder.

26. $11 \overline{)9.0}$

27. $9 \overline{)5}$

Solve for the unknown. Check your answer by using it in the original problem.

28. 3.2X + 0.07 = 4.55

29. Check for #28

Fill in the blanks.

30. A _____ has infinite length and width and is said to be two-dimensional.

31. A _____ has two endpoints.

32. A _____ has no length or width.

33. A line has _____ but no _____ .

34. A _____ has one endpoint.

35. An angle with a measure less than 180° but greater than 90° is a(n) _____ angle.

36. An angle with a measure less than 90° but greater than 0° is a(n) _____ angle.

37. Two figures that are the same shape but different sizes are said to be _____ .

38. There are _____ degrees in a circle.

39. The measure of a right angle is _____ .

40. An angle with a measure of 180° is a(n) _____ angle.

41. Two shapes that are exactly the same are said to be _____ .

42. Find the approximate area and circumference of a circle that has a radius of three feet.

43. Judith received the following amounts of money for doing chores: $5, $7, $3.50, $5, and $8. Give the mean, median, and mode for her earnings.

44. Melanie ordered books from a catalog. The prices of the books added up to $45.60. She had to pay a 6% tax and 8% for shipping. What was the total cost of her order?

45. Brandon entered a contest for free math materials. A total of 758 people each put in one entry, and there will be only one winner. What is the probability of Brandon winning the contest?